Our Adventure Vacation

by Sally Rippin
illustrated by Rob Mancini

EDUCATORS PUBLISHING SERVICE
Cambridge and Toronto

Series Authors: Kay Kovalevs and Alison Dewsbury
Series Consultant: Melinda Rice
Commissioning Editors: Sarah Russell, Bonnie Lass, and Sethany Rancier Alongi
Senior Editorial Manager: Sheila Neylon
Text by Sally Rippin
Illustrated by Rob Mancini
Edited by Anne McKenna, Laura A. Woollett, and Stacey Nichols Kim
Designed by Tanya De Silva

Making Connections program developed by Educators Publishing Service, a division of School Specialty Publishing and by Harcourt Education, a division of Reed International Books Australia Pty Ltd

ISBN 0-8388-3318-7

2010 2009 2008 2007 2006
10 9 8 7 6 5 4 3 2 1

Printed in China

Our last vacation was great. We usually spend our vacations in the city with Grandma because Dad has to work. But this time, Dad got some time off.

It didn't start out well, but it ended up being the best vacation Millie and I have ever had. Here's what happened.

We were really excited on the first morning of summer vacation. Dad was taking us fishing!

At breakfast, Dad said, "Well, Libby and Millie, are you ready to catch some fish?"

We nodded.

"Why don't you play with your cars while I get the fishing rods from the basement?" Dad said.

Millie and I set up a really good racetrack. It started on the corner of Dad's desk, ran down his chair, and did a loop on the floor.

Bang! We jumped when we heard the basement door slam.

"Dad's back," said Millie. "It doesn't sound like he found the fishing rods."

“Grrr!” Dad growled as he walked into the room. “I can’t find the fishing rods. But I wonder . . . ”

Dad was getting an idea. We held our breaths.

"We don't need to fish!" he exclaimed. "Let's go camping!"

"Camping?" Millie and I said together.

"It'll be lots of fun. You'll see," Dad said. "I saw the old tent when I was looking for the fishing rods. We'll take the subway to the dock. Then we'll get the ferry to Baker Island."

"Hooray!" Millie and I yelled.

The next day, Millie and I got up early. Dad had packed everything the night before. Soon after breakfast, we set off.

Millie and I each had a little backpack. Dad had a great big bag. He was carrying the tent, the food, and his clothes. We only had a toothbrush, our pajamas, a change of clothes, and some toy cars in our backpacks.

First, we caught the subway to the ferry dock. The ferry was waiting for us when we arrived. The water slapped against the sides of the boat.

Dad bought a ticket for each of us, and we got on board.

When we arrived at Baker Island, it was very busy. There were already lots of people there having fun.

We got on an old bus. The seats were hot and sticky and the bus drove very slowly. But we had a good time looking out the windows at all the different trees and the people riding their bikes.

We got off the bus at the state park. The park had a beach with golden sand. The water was so bright that it hurt our eyes. Dad pointed to a spot in the shade.

"That's our campsite," Dad said. "Let's set up the tent. Throw me the stakes, Libby. They're in the pocket of my bag."

I looked in the pocket of Dad's bag. No stakes. Millie and I began to get worried. Dad spent several minutes looking for the stakes until he finally gave up. He wrapped the tent around himself like a cocoon. "Maybe we should sleep like caterpillars?" he joked. Millie and I laughed.

"It doesn't matter," he said. "We'll use sticks."

We found some strong sticks in the woods. A short while later, our little tent was up.

"Let's go for a swim," Dad said.

The water was so clear we could see our toes on the bottom. I collected pink and white shells on the beach while Dad and Millie splashed in the water.

"Hey, who's hungry?" Dad yelled, running out of the water to the tent.

"I am! I am!" Millie and I yelled, running after him.

"Let's have eggs and bacon!" said Dad. He took a frying pan and a little gas burner out of his big bag.

But there were no matches!

“It doesn’t matter,” said Dad. “It’s too hot to cook anyway.”

So we ate cheese and crackers and brownies. Then we set up a racetrack for our cars with the shells we had collected.

Soon it was time for bed. It was hot in the tent, but Millie and I felt happy. We looked at the stars through the tent flap while Dad snored.

The next morning, we had peanut butter sandwiches and apples for breakfast.

"That's the end of the food," said Dad. "Everything else has to be cooked. I guess we'd better go back. Sorry, Millie and Libby, this was a disaster, wasn't it?"

“No way, Dad!” we said, hugging him. “It was the best vacation ever!”

“Really?” asked Dad.

“Yes!” I cried. “We got to go swimming and sleep in a tent.”

“And we ate cheese and crackers and brownies for dinner!” said Millie.

We packed up the tent and headed home. First, we took the old bus. Then, we got on board the ferry. When we arrived at the ferry dock, we walked to the subway stop and took the train home.

That night, we told Dad that sleeping in an ordinary bed didn't seem like fun after sleeping in the tent. So Dad helped us build a fort in the living room out of chairs and a sheet. We fell asleep in our fort and dreamed about our camping adventure on the island.